I Am a Platypus

BY Alivia Clark

ISBN: 978-1-948638-61-6

Published by

Fideli Publishing, Inc.
119 W. Morgan St.
Martinsville, IN 46151

www.FideliPublishing.com

Mr. and Mrs. Platypus woke up at the break of dawn.

The day had come when their eldest son would have to leave the pond. The rest of the platypus parents, were also wide-awake. They too were sending their children off to the school named Academy Drake.

The Mayor of the town of Drake had made a new school rule, that all children in the town would attend the exact same school.

And even though Academy Drake was a school built for ducks, the Mayor had decided, "That doesn't matter all that much. We all learn better when seated together." And that's the message he sent in his letter.

The platypus parents protested, and said they didn't agree."

"That's too bad," the Mayor said, "for that's the way it will be!"

So, finally, the day had come, the first day of school under the new rule.

So, Duckbill's mother woke him up to get him ready for school.

"Now listen here," his mother sighed, as she combed his thick fur to one side. "Be patient and don't worry if some ducks are cruel, for you will find that if you are kind, most ducks in school are cool."

Deep inside though, she felt unsure and afraid for her little Puggles, attending a school meant for ducks, she knew it would be a struggle.

"Remember our pledge," his father said, as he helped little Duckbill make his bed.

"Say it loud, say it proud, and say it from your heart. Be sure to let the teachers know that platypuses are smart!"

Duckbill smiled to himself, as he bent to tie his shoe. 'Duh!" he said to his dad, "I've known my pledge since I was two!"

They kissed good-bye, and he closed the gate, and walked one mile south to Academy Drake.

Duckbill opened the classroom door, rushed in without a peek, took off his backpack, and sat right down in the first available seat.

"Hi, my name's Duckbill," he called out to Ms. Mack.

Ms. Mack looked up from her desk and said, "Hello," right back.

The ducks stared at Duckbill, and some began to quack. He wondered why they stared at him, it made him feel on edge. "Come now," Ms. Mack told the class, "it's time to recite our pledge."

Duckbill and the ducklings stood, their eyes were on the chart. Ms. Mack flapped her wings two times, and gave the signal to start.

Duckbill knew the platypus pledge, he knew it all by heart.

But the words that he recited were not on Ms. Mack's chart.

"Let's move on said Ms. Mack. "We've finished our

reciting. Take out your paper and a pen. We're going to do some writing.

"Write about your first fly south, write about the flyway."

Duckbill had never flown before, and he'd never heard of flyway. *I know what I'll do*, he thought. *I'll write my paper my way!*

He wrote about the water games he played with his friend Guy. He wrote about eating worms, fish eggs and dragonflies.

He wrote about the funny times, his sisters played pretend.

And last, he wrote about the time, his dad spanked his rear end!

Duckbill was the first one called to read to the class.

The ducklings loved his story, and they all cheered and laughed.

All except the Buckley twins — they frowned and they stared.

They both got out of their seats and shouted, "Now just a moment here! He didn't mention flyway or write about first flight. So, stop this silly cheering, his paper isn't right."

"Let's move on now," said Ms. Mack, we have one test to go."

"I'll read the *Duck Facts* book to you, and check for what you know."

She called the ducklings one by one, to test what they had learned.

The ducks recited back the facts that make all ducklings birds:

"Our blood is warm.

"We hatch from eggs.

"We all have bills not beaks.

"But it's our feathers we all grow that make us so unique.

"We use our bill to scoop our food, whenever we want more, eating plants and insects, makes us omnivores.

"We use our bodies and our wings to take us into flight. All these things make us birds," the ducks yelled with delight!"

"Very good my little drakes another "A" you've earned."

Duckbill raised his hand to speak, he wanted to take his turn.

"Our blood is warm.

"We hatch from eggs.

"We all have bills, not beaks, and there is more than just one thing that makes us so unique.

"Our fur is thick, it keeps us warm.

"Our tails are wide and flat.

"We use our tail to help us swim, we use it to store fat.

"We use our bill to find our food, whenever we want more.

"We eat insects and not plants, so that makes us carnivores.

"We feed on our mother's milk to grow up big and strong.

"That makes us mammals and not birds, the ducks got those parts wrong!"

All the ducklings looked at him in shock. The Buckley Twins began to squawk!

"Ms. Mack," they cried in unison, "please send him out, you must! As you can see, he's something strange, he is *not* a duck like us!"

"All right now," said Ms. Mack. "Please, don't make a fuss. Our new kid here named Duckbill is a Duckbilled Platypus!"

Duckbill began to cry. His feelings were awfully hurt. He wiped his tears and snotty nose on the sleeve of his new school shirt.

Ms. Mack walked over to him, and she sat down by his side. She put her arms around him, and asked him not to cry.

"It's okay... So, you're not a duck and you have a peculiar name. What matters most is we're all here, to learn just the same."

"A platypus is what I am, and who I want to be! I'm not the same as a duck, Drake School is not for me."

The ducklings gathered around him and asked Duckbill to stay, and on the playground after lunch, they all wanted to play. All except the Buckley Twins — they frowned and stayed away.

When Duckbill got back to the pond, his parents asked, "What did you learn?"

"Well, I'm not a duck. I cannot fly. I don't have wing or feathers. I don't eat plants. I'm not a bird, and we don't fly south together.

"I do not know the school-wide pledge. The one I learned is wrong.

"The Buckley Twins would not make friends, they said I don't belong.

"I learned a lot about what I'm not on my first day at school.

"But that's not all, I also learned, most ducks in school are cool."

That night, the parents called the mayor and said, "Get rid of this new rule and let our children once again attend our platypus schools."

"I'm sorry," said the Mayor. "I made a big mistake. I'll fix it back the way it was as soon as I awake."

From that day forward, all platypus children would attend their platypus schools.

They learn their pledge and other facts that make being a platypus cool.

About the Author

Alivia Clark lives in Pittsburgh Pa. where she was born and raised. Her love for reading and learning is the result of the many books and stories her mother read to her and her nine siblings while growing up. This early love for reading influenced her career choice as an elementary school teacher and principal.

She wrote, I Am a Platypus to demonstrate the importance for students to be acknowledged and valued for the uniqueness they bring to the learning environment.

She is a proud mother of four amazing daughters and two adorable granddaughters.

Alivia Clark received her Bachelor's Degree in Elementary Education, at Carlow University and her Master's Degree in Public Management at Carnegie Mellon University.

www.ingramcontent.com/pod-product-compliance
Lightning Source LLC
Chambersburg PA
CBHW081750200326
41597CB00024B/4458

9 781948 638616